I0510227

Attain the House with All Needs

Attainable Housing Standards by Sheeraz
(The Affordable and Sustainable Housing)

Author:
Muhammad Sheeraz

International Edition

IPO (Intellectual Property Organization of Pakistan)
Trade Marks: 587956 & 584415
Copyright: 42951 - Copr
Patent: PK/P/2020/808845

Member of: REPA (Real Estate Professional Association)
www.repa.org.pk

Published
Simultaneously by:
Shamsi Publishers
36 Waqas Centre, Muhammad Bin Qasim Road,
Karachi, Pakistan Phone: +92 300 243 3716
and
Centre of Excellence for Scientific & Research Journalism
COES&RJ LLC.,
10685-B Hazelhurst Dr., Houston, TX- 77043, USA
Phone: +1-281-407-7509 Fax: +1-281-754-4941
Email: info@centreofexcellence.net
Website: www.centreofexcellence.net
Houston – London – Karachi – Singapore

Title Design by: Muhammad Sheeraz
Copyright © Muhammad Sheeraz, 2021
All rights reserved
Page 65 constitutes an extension of this copyright page.

ISBN: 978-1-716-11001-6

* Need the same book in any other language,
Please inform to Author or Publisher.

To:

Mr. Adnan Asdar Ali and Mr. Ariff Ul Islam

AAA PARTNERSHIP PVT LTD
PROJECT & CONSTRUCTION MANAGEMENT

www.aaaprojects.com

and

- Professionals
- Students
- Families who are planning to build or purchase their home.

I wish, all passionate healthy lives, easily achieve the needs, which is the biggest luxury.

DISCLAIMER

After conducting, the research and case study about house with all residential & community's needs,
Mr. Muhammad Sheeraz S/o Abdul Hameed
[M/s. Attainable Housing (Private) Limited] organized the Standards mentioned herein and named "Attainable Housing Standards by Sheeraz" are necessary to attain "The Affordable and sustainable House" which may comply all the needs.

Mr. Muhammad Sheeraz S/o Abdul Hameed
[M/s. Attainable Housing (Private) Limited] cannot guarantee that there are not mistakes or errors. These Standards not innovated with an intent to harm, injure or defame any person and not to replace or substitute for the services of trained professionals.

The professionals, students and individuals are welcome to use these Standards for the practice but should mention "Attainable Housing Standards" in the drawings and documents, which always appreciated. The Developers and anyone who will mention "Attainable Housing Standards" for the commercial purpose shall take the formal permission. Anyone who uses these Standards should always do their diligence without compromising the local and international applicable rules and regulations.

All the rights are reserved including, but not limited to update the Standards time to time.

1) Location

2) Revewable Green Energy

3) Organic Farming with Livestock

4) Community-Focused

5) Easy Payments

6) Free / Subsidized Property

7) Approvals & Ownership

8) Architectural Design with Parking

9) Design & Color Customization

10) Safety & Security

11) Construction Quality

12) Worship and Community Activities & Events

13) Affordable Education & Health Care Facility

14) Commercial Area

15) Recreational Activities

16) Green Environment / Community

17) Water Sensitive Urban Design (WSUD) with Small Dams

18) Sustainable Waste Management

19) Facility Management

"Attainable Housing Standards by Sheeraz"
(The Affordable and Sustainable Housing)

Revision-00 (December 2020)

Organized by:

Mr. Muhammad Sheeraz S/o Abdul Hameed

The Affordable and Sustainable Housing

www.ahspakistan.info

INDEX

IMPROTANT NOTES

- **"Attainable Housing Standards by Sheeraz** *(The Affordable and Sustainable Housing)*" shall be applied without compromising local regulations, applicable standards and codes.

- **"Attainable Housing Standards by Sheeraz** *(The Affordable and Sustainable Housing)*" shall be used for each type (small, medium and large size) individual house, individual farmhouse, housing society project, farmhouse projects with urban planning (horizontal development) and residential building project "mixed-use" (vertical development). Some standards to be exempted for individual house, farmhouse and vertical development, for new projects and for existing projects. Existing project by its improvement and implementation of the standards.

- **"Attainable Housing Standards by Sheeraz** *(The Affordable and Sustainable Housing)*" shall be used for "M/s Attainable Housing (Private) Limited" own projects and for other organization, individual (s), developer's new projects, and for their existing projects for sustainable development.

- **"Attainable Housing Standards by Sheeraz** *(The Affordable and Sustainable Housing)***"** are allowed free of cost to educational institutes for education purpose, students for their study purpose, professionals for their practice except, developers and any specific person / department / institute / organization who shall be allowed against fee by M/s Attainable Housing (Private) Limited.

- **"Attainable Housing Standards by Sheeraz** *(The Affordable and Sustainable Housing)***"** shall be evaluated and certified by M/s Attainable Housing (Private) Limited for commercial and other purpose against fee.

INTRODUCTION

Affordable & Sustainable Housing are global needs for all communities & nations. Attainable Housing Standards by Sheeraz set out all the needs by implementing 19 Standards. These standards to be applied without compromising local regulations, applicable standards and codes. The purpose of these standards to build a new community or to improve the existing, without destroying the ecosystem and harming the environment, and community health. Attainable housing standards are:

1) Location

2) Renewable Green Energy

3) Organic Farming with Livestock

4) Community-Focused

5) Easy Payments

6) Free / Subsidized Property

7) Approvals & Ownership

8) Architectural Design with Parking

9) Design & Color Customization

10) Safety & Security

11) Construction Quality

12) Worship and Community Activities & Events

13) Affordable Education & Health Care Facility

14) Commercial Area

15) Recreational Activities

16) Green Environment / Community

17) Water Sensitive Urban Design (WSUD) with Small Dams

18) Sustainable Waste Management

19) Facility Management

Each standard contains inputs and processes for consideration and implementation but not limited to the inputs and processes specified in the section of detailed specification. **"Attainable Housing Standards by Sheeraz** *(The Affordable and Sustainable Housing)***"** ensure that The Affordable and Sustainable Housing is attainable for all by comforts, own renewable green energy, savings, health by green community and organic farming with livestock within the project, safety & security, desirability, required amenities and above all with affordability.

PURPOSE

Living and eating are every one's basic needs. The Affordable Housing with appropriate affordable energy (electric & gas), healthy food, water and other necessary facilities are highly important for all countries and nations.

Housing with the typical and traditional amenities are not enough and needs further improvement by controlling the energy crises with the green environment, providing healthy food, disaster and waste management with the desirable home design options. Energy using oil and gas, effects the environment for present and future generation as well. The food (milk, fruit, vegetable & meat) mostly are being produced using chemical that are harmful for the health and weak the immune system, which fights against the disease, bacteria and virus. Parallel with these matters the housing prices are also becoming unaffordable even without focusing on the problems mentioned earlier.

United Nations Member States adopted The Sustainable Development Goals (17 SDGs) in 2015. The 17 SDGs are integrated that development must balance social, economic and environmental sustainability. UNDP supports countries in achieving the SDGs through integrated solutions.

"Attainable Housing Standards by Sheeraz *(The Affordable and Sustainable Housing)*" relates generally to the Housing and particularly to the Housing Standards without compromising local regulations, applicable standards and codes. The standards are for the design and performance purpose for The Affordable and Sustainable Housing.

SUMMARY OF STANDARDS

"Attainable Housing Standards by Sheeraz *(The Affordable and Sustainable Housing)*" containing 19 Standards, those must be considered and implemented by the inputs and processes specified in detailed specification. These standards will help to resolve the problems, develop a balance social, economic and environmental sustainability by providing clean renewable green energy, healthy organic food & other organic natural products availability, user desirability (in terms of home design and options) and address other necessary issues in the sustainable manner and lastly to make them affordable for all.

The standards shall be applied without compromising local regulations and applicable standards. The standards are for for each type (small, medium and large size) individual house, individual farmhouse, housing society project, farmhouse projects with urban planning (horizontal development) and residential building project "mixed-use" (vertical development). Some standards to be exempted for individual house, farmhouse and vertical development.

The standards have organized for new projects and for existing projects. Existing project by its improvement and implementation of the standards.The standards are for "M/s Attainable Housing (Private) Limited" own projects and for other organization, individual (s), developer's new projects, and for their existing projects for sustainable development.

The standards to be allowed free of cost to educational institutes for education purpose, students for their study purpose, professionals for their practice except, developers and any specific person / department / institute / organization who shall be allowed against fee by M/s Attainable Housing (Private) Limited. The standards shall also be evaluated and certified by M/s Attainable Housing (Private) Limited for commercial and other purpose against fee.

Note: The amended as necessary and based on feedback from assessors and industry.

HOW TO USE STANDARDS

The standards are containing 19 Standards for The Affordable and Sustainable Housing.

19 Standards:
19 Standards that are to be considered and implemented by inputs and processes

Inputs and Processes for consideration and implementation:
Standards are containing information to implement the standards. The detail and example to implement the standards in standard layout and proposed floor plans are not meant to be exhaustive or prescriptive but rather to provide key points and guidance in how to achieve the standards. There are a number of inputs and process in each standard for good practice that also provide the guidance about how to achieve the standards, without being exhaustive or prescriptive. Housing society projects and building projects might do some of these things, but might also be doing a range of different things that equally demonstrate that they meet the standard. Implementation of the standards and evaluation shall be based on how things are actually done and what is achieved, not just on what is written in detailed description of "**Attainable Housing Standards by Sheeraz** *(The Affordable and Sustainable Housing)*".

DETAILED DESCRIPTON OF STANDARDS

The affordability and sustainability will be evaluated against 19 Standards. Detailed descriptions and specifications for each standard are:

1 - Location

Purpose	Purpose of this standard is to consider and implement the inputs and processes for the project location and location of houses within the project. The project (housing society or building project) location within the city or near the city always easy to access using public transport. Appropriate location of the house and amenities in the housing society helps to receive natural energy and to improve sustainable living by walking instead of using vehicle from one space to the other space within the project.
AHS-1.1: 2020	Local transport facility within 800 meters distance from the project to access the city. (if project is not within the city) *
AHS-1.2: 2020	Easy access to amenities for elderly, children and disabled.
AHS-1.3: 2020	Amenities & commercial area within 600 meters distance from the residence. Lift facility for more than 3 floors in building project.

1 - Location (Continue......)

AHS-1.4: 2020	Proper signage system for help and guidance of destinations.
AHS-1.5: 2020	Allocation of houses / apartments as per buyer's choice on first-come-first-serve basis.

Note: AHS-1.1: 2020 "Local city transport may be provided by Local Authorities or by any private sector or by the developer to comply the standard"

2 - Renewable Green Energy

Purpose	Purpose of this standard is to consider and implement renewable green energy options for residence, commercial areas and other needs of the housing society. Renewable green energy is good for planet and for the people. It provides reliable energy in terms of Electric Power and GAS. It is important to prevent shortage. It reduces regular energy cost, enhance energy security, provides fuel diversification and helps to conserve the nation's natural resources. Renewable green energy emits no or low "greenhouse gases" and "air pollutants".
AHS-2.1: 2020	Centralized or Decentralized energy by solar and / or wind turbine system to each residence, commercial area and for amenities. Decentralized is recommended as individual for each house.
AHS-2.2: 2020	Solar and / or wind turbine system for streetlights, parks, livestock and other areas.
AHS-2.3: 2020	Centralized Biogas plant with livestock for the residence for Gas supply.
AHS-2.4: 2020	Solar heater for residence and amenities.

2 - Renewable Green Energy (Continue......)

AHS-2.5: 2020	Considering and implementation of any other renewable low cost green energy system such as energy from waste etc. without harming the environment & health. Decentralized system (individual for each house) to be preferred.

Note: On-grid power supply and locally available Gas supply lines as primary source may remain connected or terminated, subject to authority approvals, building regulations and fulfil of energy requirements by renewable green source.

3 - Organic Farming with Livestock

Purpose	Purpose of this standard is to consider and implement the inputs and processes to achieve organic food and other organic natural products. Pure milk, healthy fruits, vegetables and many other daily use products are essential for everyone. This is necessary to make them attainable for all. Organic foods (free from pesticides) always better and not harmful for health. Organic and natural foods provide energy, nutrients, vitamins and boost immune system for fight against dieses, bacteria and viruses. The livestock is not only necessary for organic food production but itself also provides meat and other products.
AHS-3.1: 2020	Organic farm for different fruits, vegetables and crops with livestock for the residents. Forest farming with boundary wall of the society, centralized organic farm and livestock farming, using approx. 10% of total land in housing society. *(refer to Drawing Sheet No. 001 - AHS Standard Main Layout)* The building project may apply this standard for organic farming far from the project by third party service.

3 - Organic Farming with Livestock
(Continue......)

AHS-3.2: 2020	Livestock with organic farm for pure milk and other products as well as for organic fertilizer. *(refer to Drawing Sheet No. 001 - AHS Standard Main Layout)*
AHS-3.3: 2020	Fishpond in central park connected with centralized organic farm & livestock. Fishpond may be managed in small dam. *(refer to Drawing Sheet No. 001 - AHS Standard Main Layout)*
AHS-3.4: 2020	Organic farming with honeybee for natural honey production with safety measures against honeybee attack.
AHS-3.5: 2020	Organic farming for skincare and functional cosmetics products.

Note: Organic farming and livestock far from the project, at the other location for organic food and organic natural products to be provided by third party but production and operations for standards implementation will be treated the same way as farming within the project.

4 - Community-Focused

Purpose	Purpose of this standard is to consider and implement the inputs and processes to focus on the communities / overseas, community needs and the culture. Focus on the needs and the culture of any specific community / overseas and the families who intend to build many houses at the same location, is essential for a happy life. It is an emphasis on participant's ideas, concerns, and interactions.
AHS-4.1: 2020	Focused on the communities and the culture by providing their additional boundary. *(refer to Drawing Sheet No. 001 - AHS Standard Main Layout)*
AHS-4.2: 2020	External private open space for specific community and family.
AHS-4.3: 2020	Small-scale organic farming as kitchen grading with small fishpond in community system. *(refer to Drawing Sheet No. 001 - AHS Standard Main Layout)*
AHS-4.4: 2020	Additional amenities as per community culture. (scale of amenities is subject to scale of project and residence quantity of specific community)

4 - Community-Focused (Continue......)

AHS-4.5: 2020	Houses and residential units in community system may include additional boundary, shared spaces concept such as shared large kitchen, shared dining area, and shared laundry, and small scale recreational activities. This type of community to be designed specially as desired by the community.

5 - Easy Payments

Purpose	Purpose of this standard is to consider and implement the inputs and processes to make affordable price and easy payment procedures. Considering all appropriate available options to reduce the cost of property and affordable payment plan are necessary to make the property attainable for all.
AHS-5.1: 2020	Review and implement all appropriate available options to waive or reduce the bank's EMIs (interest burden) by support; from the government, financial institutes, any organization (NGO) or individual (s).
AHS-5.2: 2020	Affordable payment plan as per buyer's comfortable schedule.
AHS-5.3: 2020	Proper construction-linked payment plan without any type of interest amount by the developer in case of payment delay from the buyer.

6 - Free / Subsidized Property

Purpose	Purpose of this standard is to consider and implement the inputs and processes to provide free and affordable houses and property by subsidy, for the people who do not have a lot of money. The aimed towards alleviating housing costs and expenses for impoverished people with low to moderate incomes.
AHS-6.1: 2020	Subsidized affordable prices for people with low to moderate incomes. Subsidy by the government, financial institute, any organization (NGO), individual (s) and by any specific community for their own people and others.
AHS-6.2: 2020	Subsidized rent after purchasing the property by the government, financial institute, any organization (NGO), individual (s) and by any specific community for their own people and others, to overcome the poverty.
AHS-6.3: 2020	Free or subsidized houses and property option by transparently lucky draw (if feasible)

Note: Subsidized house and property could be offered based on affordability, not on the size or type of house. The subsidized rent could be determined by family income, which called "rent-geared-to-income" house.

7 - Approvals & Ownership

Purpose	Purpose of this standard is to consider and implement the inputs and processes to make ensure the authority approvals and ownership of the property buyer. This includes the property land title that effect the property current price & future price, and associated risks (if / any) in future. Even the authority approval is sufficient but ownership of the buyer is also matter for future purpose.
AHS-7.1: 2020	Proper approvals from local authorities and necessary NOCs from all departments must be obtained.
AHS-7.2: 2020	Authorities approvals for houses may be obtained with complete housing project as to manage compulsory open space (COS) with the main layout. *(refer to Drawing Sheet No. 001 - AHS Standard Main Layout and AHS proposed layouts of each type house; drawing Sheet No.: A-120, B-120 U-120, A-85, B-85 and U-85)*
AHS-7.3: 2020	House and property buyers must have complete ownership of their property as per terms and conditions set out by the developers.

8 - Architectural Design with Parking

Purpose	Purpose of this standard is to consider and implement the inputs and processes related to architectural designing of the house and housing society / building with the parking facility for each house. Architectural design to consider the residence needs without compromising the applicable local land-use codes and design standards. It is also necessary to qualify affordable and sustainable housing a project should include a range of housing types. The provision of various designs of houses, housing units and buildings can support a more diverse population and allow more equitable distribution of households of all income levels.
AHS-8.1: 2020	Orientation of house and apartment shall be maximum as per regionally confirmed wind direction to receive maximum fresh air.
AHS-8.2: 2020	Width of Bedroom minimum 9'-0" (2.7 meters) however, recommended 10'-0" (3 meters). One bathroom including W/C and shower for two bedroom is adequate but separate bathroom is recommended for each bedroom. *(refer to AHS proposed layouts of each type house; drawing Sheet No.: A-120, B-120 U-120, A-85, B-85 and U-85)*

8 - Architectural Design with Parking
(Continue......)

AHS-8.3: 2020	Kitchen must have exhaust system and one window directly to open area is to be preferred. Allowance for kitchen cabinets for storage space. *(refer to AHS proposed layouts of each type house; drawing Sheet No.: A-120, B-120 U-120, A-85, B-85 and U-85)*
AHS-8.4: 2020	Appropriate laundry area / washing machine space with water supply and drainage lines. *(refer to AHS proposed layouts of each type house; drawing Sheet No.: A-120, B-120 U-120, A-85, B-85 and U-85)*
AHS-8.5: 2020	Appropriate location for water pump machine.
AHS-8.6: 2020	Space provision for generator if required by the resident.
AHS-8.7: 2020	Water supply for hot and cold water for bath & kitchen.
AHS-8.8: 2020	Consider the indoor air quality and daylight
AHS-8.9: 2020	Acoustic and thermal material to be considered and enhanced.
AHS-8.10: 2020	Introduce low energy and water efficient home appliances to the residents, from initial design and construction stage and develop awareness for it by literature, information and organizing community events.

8 - Architectural Design with Parking
(Continue......)

AHS-8.11: 2020	House roof covered by solar panel and remaining are for roof top kitchen garden. Rooftop, Terrance and balconies must have appropriate draining system and adequate slope for water runoff.
AHS-8.12: 2020	Planter must be considered in terrace, balconies, stairs landing and all possible space and grow the plant if kitchen gardening not possible so the plants and trees to be considered for wall covering and to develop the possible green house without compromising the local codes and applicable architectural standards.
AHS-8.13: 2020	Staircase should comply with the building regulations and must be enough wide to shift the wheelchair households. It is recommended that staircase for residential units where one staircase will be used for approximately 6 houses may not be less then minimum 4'-0" width (1.2 meter). *(refer to AHS proposed layouts of each type house; drawing Sheet No.: A-120, B-120 U-120, A-85, B-85 and U-85)*

8 - Architectural Design with Parking
(Continue......)

AHS-8.14: 2020	Parking for each house must be accommodated. Parking may be designed partially or complete inside / outside the house or near the house but not more than 100 meters distance from the residents. House layout may be considered conjunction with main housing layout to obtain the approval in case of small size plot, due to achieve compulsory open space (COS) regulation. *(refer to Drawing Sheet No. 001 - AHS Standard Main Layout and AHS proposed layouts of each type house; drawing Sheet No.: A-120, B-120 U-120, A-85, B-85 and U-85)*
AHS-8.15: 2020	Sunshade to be designed to control the direct sun heat and rainwater.

Note: The same standards must be considered for Education, health care building and other amenities.

9 - Design & Color Customization

Purpose	Purpose of this standard is to consider and implement the inputs and processes about house customization in terms of architecture amendments and desire color. Customization option can full fill the more requirements of residents and make the house feel like home. The primary use of this standard is for the developers where the buyer can began their house customization with predefined architecture layout, if allowable by the developer's as per described terms and conditions. The below to be consider:
AHS-9.1: 2020	Apply the resident's preferred internal and external (elevation) color shade for the house as per terms and condition set out by the developers. Apartments and residential units may have color customization for interior and may not be allowable for exterior (elevation).
AHS-9.2: 2020	Amendment in architectural design without structure alteration as per terms and condition set out by the developers.
AHS-9.3: 2020	Provision for circulation and functionality to meet the needs of wheelchair households including one handicapped toilet. It will be not less than local authority standards and any applicable codes. This may be part of initial client brief and to be incorporated in architecture design accordingly, if required by the residents.

10 - Safety & Security

Purpose	Purpose of this standard is to consider and implement the inputs and processes related to all safety & security measures of residence, their house and complete housing society / building project. The housing safety and security hazards assessment and controlling to be focused for the houses and public places.
AHS-10.1: 2020	Security cameras at the entry points, specific boundary locations and each sensitive location of the housing society and building.
AHS-10.2: 2020	Light up the landscape and hiding places in the society / building.
AHS-10.3: 2020	Proper road marking and appropriate road crossing in housing society project. *(refer to Drawing Sheet No. 001 - AHS Standard Main Layout)*
AHS-10.4: 2020	Children safety from windows, balcony and roof parapet for which safety grill to be installed. Safety measures must be applied as per local authority's regulations and applicable safety codes. The parapet recommended height is 3'-10" (1.2 meter) from finish floor level if not specified in regulations and applicable safety codes.
AHS-10.5: 2020	Safety devices for electrical hazards to be installed in each house / apartment.

10 - Safety & Security (Continue......)

AHS-10.6: 2020	Fire extinguishers as fire point at corner of each street in housing society and in the corridor at each floor in building project. *(refer to Drawing Sheet No. 001 - AHS Standard Main Layout)*
AHS-10.7: 2020	Fire compartmentalization and fire exit with adequate fire resistance-rated material, Exit signs, Fire alarm system, Sprinkler system but, affordability and sustainability must be considered.

Note: The same standards must be considered for Education, health care building and other amenities. Any other / additional safety & security hazards assessment and controlling will be appreciated in standards assessment and the same may be considered the part of standards in future. However, affordability and sustainability must be measured.

11 - Construction Quality

Purpose	Purpose of this standard is to consider and implement the inputs and processes related to construction quality. Ensure the quality delivery, review the results and make improvements as required are the part of this standard. Quality in construction means that a work is complete within the defined guidelines to avoid rework and improve construction life. Poor quality can cause a higher number of health and safety incident and increase running cost due to heavy maintenance.
AHS-11.1: 2020	Focus on green construction material and process, which does not harm the environment and community health.
AHS-11.2: 2020	Focus on wall, roof and other components of the envelope to make an energy efficient and manage indoor environment by considering "R Value" and other measurements.
AHS-11.3: 2020	Concrete, metal and other appropriate material to be used as per typically used in region & specifically approved for the project, as per resident's acceptance but, affordability and sustainability must be measured. (Compliance of any other applicable and appropriate codes to achieve energy efficiency shall be appreciated)

11 - Construction Quality (Continue......)

AHS-11.4: 2020	Tested concrete, specified strength as per structure calculation.
AHS-11.5: 2020	Tested cement concrete block, specified strength as per structure calculation.
AHS-11.6: 2020	Pressure tested plumbing fittings (all supply, drain and waste)
AHS-11.7: 2020	Water proofing application and testing prior to finishing for bath and other wet area.
AHS-11.8: 2020	Wooden door in regular wet area (particularly washroom) must be cladded with metal sheet to prevent damaged cause of water.
AHS-11.9: 2020	Installation of Copper wire for electric.
AHS-11.10: 2020	Power socket for appliances must be separate in power distribution board for purpose of safety and generator / UPS provision.
AHS-11.11: 2020	Earthing system with Electrical works must be provided for safety function.

12 - Worship and Community Activities & Events

Purpose	Purpose of this standard is to consider and implement the inputs and processes about establishment and maintenance of the relation between human beings with many facets from the holy to social development. Collective worship and community activities & events on daily, weekly and occasionally basis to develop connections, mobilize the people and resources to resist, reduce and end social and economic inequality. Community center with recreational & various cultural activities, social support, public information and other purpose for the whole housing society, for a specialized group or community and for specified family for their events. These are also important to engage the community to achieve long-term and sustainable outcomes, relationship and decision-making.
AHS-12.1: 2020	Location of worship building (Jama Masjid) and community center for community activities and events within 600 meters distance from residents with facing wide road, easy access and will limit the use of vehicle to decrease gas emissions further avoid to environmental pollution. *(refer to Drawing Sheet No. 001 - AHS Standard Main Layout)*

12 - Worship and Community Activities & Events (Continue......)

AHS-12.2: 2020	Additional parking facility may be provided (as much possible as per the scale of project) with worship block (Jama Masjid) and community center.
AHS-12.3: 2020	Health club / fitness center, attached with the community center or detached at the nearest of community center. *(refer to Drawing Sheet No. 001 - AHS Standard Main Layout)*

13 - Affordable Education & Health Care Facility

Purpose	Purpose of this standard is to consider and implement the inputs and processes to provide affordable education and health care facility options to develop a welfare state. Affordable education provides education opportunity for all, which develop the society for living morally, creatively, and productively. Affordable health facility always considered as basic need for all.
AHS-13.1: 2020	Education and health facility within 600 meters distance from the residents with facing wide road, easy access and will limit the use of vehicle to decrease gas emissions further avoid to environmental pollution.
AHS-13.2: 2020	Additional parking facility may be provided (as much possible as per the scale of project) with education / school and health care facility / hospital.
AHS-13.3: 2020	Park or ground facility (as much possible as per the scale of project) must be provided with education facility / school with fencing protection that may also be used by residents in school off time.

13 - Affordable Education & Health Care Facility (Continue......)

AHS-13.4: 2020	Free or affordable education facility / school must be provided with compliance of local authority's requirement, codes & regulations by considering minimum one of the following standard (other typical education facilities / school could be provided as well): i) Free education facility / school for all, operated by the government, local authorities or collectively by government & non-government resources. ii) Free education facility / school for all, operated by any welfare education organization. iii) Free or affordable education facility / school operated by the developer or any education organization but shall be monitored by the housing society. Welfare supports must be provided for all eligible students.

13 - Affordable Education & Health Care Facility (Continue......)

AHS-13.5: 2020	Appropriate and free or affordable health care facility / hospital must be provided with compliance of local authority's requirement, codes & regulations by considering minimum one of the following standard (other typical health care facilities / hospital could be provided as well): i) Free health care facility / hospital for all patients, operated by the government, local authorities or collectively by government & non-government resources. ii) Free health care facility / hospital for all patients, operated by any welfare health organization. iii) Free or affordable health care facility / hospital operated by the developer or any health organization but shall be monitored by the housing society. Welfare supports must be provided for all eligible patients.

13 - Affordable Education & Health Care Facility (Continue......)

Note: Construction and development is primary responsibility of housing developer however, full or partial support could be provided by the government, any organization and community.

Welfare support for operations may be provided by; government, local authorities, any welfare organization, NGO, individual (s), collectively by different organizations and communities or by different communities for their specified eligible students and patients. Education facility and health care facility shall be only for housing residents except otherwise permitted by the developer based on resident's voting in which number of houses to be calculated and, parking facility must be considered accordingly.

14 - Commercial Area

Purpose	Purpose of this standard is to consider and implement the inputs and processes specifically about commercial area for use by for-profit business, where goods or services will be bought, sold or exchanged. Dedicated commercial area with market zoning concept to provide easy way to find required goods and services. This standards also focus on not to design and execute commercial activities in residential area except otherwise designed as cultural based block where specified block with additional boundary / fence to be designed for commercial and residential mixed-use at the same plot.
AHS-14.1: 2020	Commercial area preferred more central locations or within 600 meters distance from the residents with facing wide road, easy access and will limit the use of vehicle to decrease gas emissions further avoid to environmental pollution. More than one commercial area is preferred for large-scale housing society. *(refer to Drawing Sheet No. 001 - AHS Standard Main Layout)*

14 - Commercial Area (Continue......)

AHS-14.2: 2020	Commercial area must have market zoning where the same type of goods and services will be at the same location such as meat, vegetable & fruit which are daily kitchen needs shall be at one location and refreshment, milk, bakery at other location. Large project can have more market zones such as separate zone for meat, separate zone for vegetable and the same for other goods and services.
AHS-14.3: 2020	Stalls facility as small and temporary sale point options must be considered subject to housing society / building project design and concept, and requirements of the community and location as per approval by the local authorities and as per terms and conditions set out by the developers.

15 - Recreational Activities

Purpose	Purpose of this standard is to consider and implement the inputs and processes related to recreational activities, which play an important role in communities such as improving the health and promoting the development of communities. Leisure activities like fishing, outdoor physical and social activities, and recreational activities like golf, recreational hobbies like gardening are part of this standard.
AHS-15.1: 2020	Park in each street with mini golf course (if possible) for outdoor recreational physical activities and kitchen gardening (if possible) as recreational hobbies. *(refer to Drawing Sheet No. 001 - AHS Standard Main Layout)*
AHS-15.2: 2020	Minimum one central park with children swings, mini zoo, jogging track and fishpond. Small dam could be managed as fishpond. *(refer to Drawing Sheet No. 001 - AHS Standard Main Layout)*
AHS-15.3: 2020	Separate parks with children swings, mini golf course, kitchen gardening and small fishpond with additional boundary for community system. *(refer to Drawing Sheet No. 001 - AHS Standard Main Layout)*

Note: Playground and other spaces for recreational activities and scale of these recreational activities will be increased and improved subject to size and type of the project.

16 - Green Environment / Community

Purpose	Purpose of this standard is to consider and implement the inputs and processes to develop green environment / community by planting the trees and shrubs with houses. Plantation that is important to control air pollution, which causes asthma and other health hazards. Trees will absorb pollutant gases and act as clean the air we breathe and filter the water community drink. Trees can help to cool down the streets, combat global warming, contribute to soil health, retain water and providing shade to the community.
AHS-16.1: 2020	Minimum 12'-0" (3.5 meters) to 15'-0" (4.5 meters) area with the boundary in housing society to be set for forest farming with different trees. *(refer to Drawing Sheet No. 001 - AHS Standard Main Layout)*
AHS-16.2: 2020	Plantation of minimum 1 tree of any size for each house and recommended 1 tree of any size for each residents for which 5 trees may be considered for each family.
AHS-16.3: 2020	Develop the system to provide awareness of green environment / community by activities and community events and provide the system to support the community for home plantation and organic kitchen gardening.

17 - Water Sensitive Urban Design (WSUD) with Small Dams

Purpose	Purpose of this standard is to consider and implement the inputs and processes to develop cost effective "Water Sensitive Urban Design (WSUD) and to focus on water supply & demand. Create wetlands options to recharge ground water. Restore storm water with adequate size of Dams options to decrease flood risk, greater security of water supply, overcome domestic water problems as well as to fulfill the landscape water demand. Reduce domestic water demands, incorporating the water efficient appliances and plumbing fittings. Build awareness and encourage the community about water conservation.
AHS-17.1: 2020	Consider water efficient shower fixture, toilet flush system and try to install toilet dual-flush system.
AHS-17.2: 2020	Water efficient landscaping to reduce potable water consumption
AHS-17.3: 2020	Small fishponds in community system and other appropriate locations with safety measures for children, as sedimentation basins. *(refer to Drawing Sheet No. 001 - AHS Standard Main Layout)*
AHS-17.4: 2020	Use maximum permeable surfaces instead of standard paving.

17 - Water Sensitive Urban Design (WSUD) with Small Dams (Continue......)

AHS-17.5: 2020	Paving with voids around, to reduce surface run-off water and allow it to drain through voids and sub-base.
AHS-17.6: 2020	Swales and buffer strips with roads and corner of streets to rainfall runoff and connected with adequate size of Dam. Bioretention system with swales or basin may be considered for road layout and streetscape.
AHS-17.7: 2020	Infiltration trenches and system with some modification for road layout and streetscape.
AHS-17.8: 2020	Rain garden around the boundary and corner of streets to retain the rain, which let the plants and trees grow.
AHS-17.9: 2020	Develop the system to recycle of used (grey) water for the landscaping, green community, organic farming, improve ecosystem health and other appropriate use.

18 - Sustainable Waste Management

Purpose	Purpose of this standard is to consider and implement the inputs and processes to develop methods to reduce the amount of natural consumed during the construction activity and after construction from the residents from their daily activities. It is highly lucrative, keeps the environment clean and fresh, saved each and conserves energy, reduces environmental pollution and highly important to develop and maintain a clean sustainable green community.
AHS-18.1: 2020	Preparation and implementation of appropriate Site Waste Management Plan (SWMP) for construction activities. Focus on construction waste by considering five R's of Waste management: reduce, reuse, recycle, recover and residual management.
AHS-18.2: 2020	Placing and managing separate waste bins according to sustainable waste management including kitchen waste bin for organic compost, recycle material waste bin, reusable material waste bin and waste bin for residual disposal. *(refer to Drawing Sheet No. 001 - AHS Standard Main Layout)*
AHS-18.3: 2020	Allocate appropriate space and manage waste compost for organic forming.

18 - Sustainable Waste Management
(Continue......)

AHS-18.4: 2020	Develop a system for residual disposal of waste by close connection with the local authority's system.
AHS-18.5: 2020	Create awareness in residents and community to collect and donate reusable material to achieve sustainable waste management goals.
AHS-18.6: 2020	Create an appropriate space to collect and manage the unwanted but reusable cloths, books, toys and other stuff form the residents for school, hospital and other needy families to reuse the waste.
AHS-18.7: 2020	Develop a system for residents to reduce the waste including focusing on reusable shopping bag instead of paper and plastic bag from the shop and prohibit plastic shopping bags as much as possible. Create awareness about the system by community events.

Note: Finding and implementation the options to reuse as many times as possible the material that are taken from nature, to manage our waste sustainably following the five R's of Waste management: reduce, reuse, recycle, recover and residual management. It will include the opportunity to achieve renewable green energy and other resources using waste. These all will be appreciated in standards assessment and the same may be considered the part of standards in future. However, affordability and sustainability must be measured.

19 - Facility Management

Purpose	Purpose of this standard is to consider and implement the inputs and processes to develop operation & maintenance system and upgrading the house and housing society / housing building project, which ensure accommodation quality and overall value sustained by great listening and empathizing.
AHS-19.1: 2020	Manage all commissioning and maintenance documents related to mechanical, electrical, plumbing and other works and ensure regular maintenance according to appropriate facility management system.
AHS-19.2: 2020	Appropriate facility management system by the developer or any other facility management firm to ensure the quality of life with peace of mind for each residents and for all public spaces.
AHS-19.3: 2020	Supervising multi-disciplinary teams of staff including cleaning, safety & security, transportation (if applicable) etc.
AHS-19.4: 2020	Ensure the basic facility such as water, electric and gas, are well maintained for each house and public space.

19 - Facility Management (Continue......)

AHS-19.5: 2020	Ensure by periodically survey and feedback that maintenance and services of all facilities meet the needs of the residents.

Note: Any additional work or procedure of facility management will be appreciated in standards assessment and the same may be considered the part of standards in future. However, affordability and sustainability must be measured.

REFERENCE DRAWINGS

Standard	Input and process
REFERENCE DRAWING: Drawing Sheet No.: 001 AHS Standard Main Layout REFERRED TO:	
3 - Organic Farming with Livestock	AHS-3.1: 2020 AHS-3.2: 2020 AHS-3.3: 2020
4 - Community-Focused	AHS-4.1: 2020 AHS-4.3: 2020
7 - Approvals & Ownership	AHS-7.2: 2020
8 - Architectural Design with Parking	AHS-8.14: 2020
10 - Safety & Security	AHS-10.3: 2020 AHS-10.6: 2020
12 - Worship and Community Activities & Events	AHS-12.1: 2020 AHS-12.3: 2020
14 - Commercial Area	AHS-14.1: 2020
15 - Recreational Activities	AHS-15.1: 2020 AHS-15.2: 2020 AHS-15.3: 2020
16 - Green Environment / Community	AHS-16.1: 2020
17 - Water Sensitive Urban Design (WSUD) with Small Dams	AHS-17.3: 2020
18 - Sustainable Waste Management	AHS-18.2: 2020

Standard	Input and process
REFERENCE DRAWING: Proposed House Layouts Drawing Sheet No.: A-120, B-120 U-120, A-85, B-85 and U-85 REFERRED TO:	
7 - Approvals & Ownership	AHS-7.2: 2020
8 - Architectural Design with Parking	AHS-8.2: 2020 AHS-8.3: 2020 AHS-8.4: 2020 AHS-8.13: 2020 AHS-8.14: 2020

THE STANDARDS RATING SYSTEM

"Attainable Housing Standards by Sheeraz *(The Affordable and Sustainable Housing)*" rating system indicated by levels according to design and performance. The rating from Level-1 to Level-3 depending on the achieved standards. Level-1 which is entry level, is completely flexible for the developers to implement. The further achievement within the level to be marked " ★ " additionally. Each, horizontal development and vertical development has different flexibility.

S. No.	Focused Issues	STANDARDS	FLEXIBILITY OF STANDARDS	
			HORIZONTAL DEVELOPMENT (HOUSING SOCIETY PROJECT)	VERTICAL DEVELOPMENT (BUILDING PROJECT)
1	Affordable & Sustainable	Location	*Compulsory at each level*	*Compulsory at each level*
2		Renewable Green Energy	*Compulsory at each level*	*Compulsory at each level*
3		Organic Farming with Livestock	*Compulsory at each level*	*Preferred at each level*
4		Community-Focused	*Preferred at each level*	*Preferred at each level*
5	Affordability	Easy Payments	*Compulsory at each level*	*Compulsory at each level*
6		Free / Subsidized Property	*Not Compulsory at Level-1*	*Not Compulsory at Level-1*
7	Desirability	Approvals & Ownership	*Compulsory at each level*	*Compulsory at each level*
8		Architectural Design with Parking	*Compulsory at each level*	*Compulsory at each level*
9		Design & Color Customization	*Not Compulsory at Level-1*	*Not Compulsory at Level-1*
10	Facilities & Affordability	Safety & Security	*Compulsory at each level*	*Compulsory at each level*
11		Construction Quality	*Compulsory at each level*	*Compulsory at each level*
12		Worship and Community Activities & Events	*Compulsory at each level*	*Preferred at each level*
13		Affordable Education & Health Care Facility	*Compulsory at each level*	*Preferred at each level*
14		Commercial Area	*Compulsory at each level*	*Not Compulsory at Level-1*
15		Recreational Activities	*Compulsory at each level*	*Preferred at each level*
16	Sustainability	Green Environment / Community	*Compulsory at each level*	*Preferred at each level*
17		Water Sensitive Urban Design (WSUD) with Small Dams	*Compulsory at each level*	*Preferred at each level*
18		Sustainable Waste Management	*Compulsory at each level*	*Compulsory at each level*
19		Facility Management	*Compulsory at each level*	*Compulsory at each level*

RATING ASSESSING SYSTEM

M/s Attainable Housing (Private) Limited shall design and develop its own projects as per **"Attainable Housing Standards by Sheeraz *(The Affordable and Sustainable Housing)*"** but not below the minimum Level-1 "Silver" mentioned herein. Other projects to be assessed as per applicable charges by M/s Attainable Housing (Private) Limited by specifically trained and accredited independent assessors.

Assessors shall conduct an initial design stage assessment, recommend a level and issue an interim certificate. A final certificate of compliance is to be issued after post-completion with two years validity. Twice a year project's random visit to be performed for assessment. The report based on visit to be prepared and incase of below standards comments, the Non Compliance Report "NCR" to be issued with deadline to close the "NCR".

ATTAINABLE HOUSING STANDARD LAYOUT
(DRAWING SHEET No.: 001)

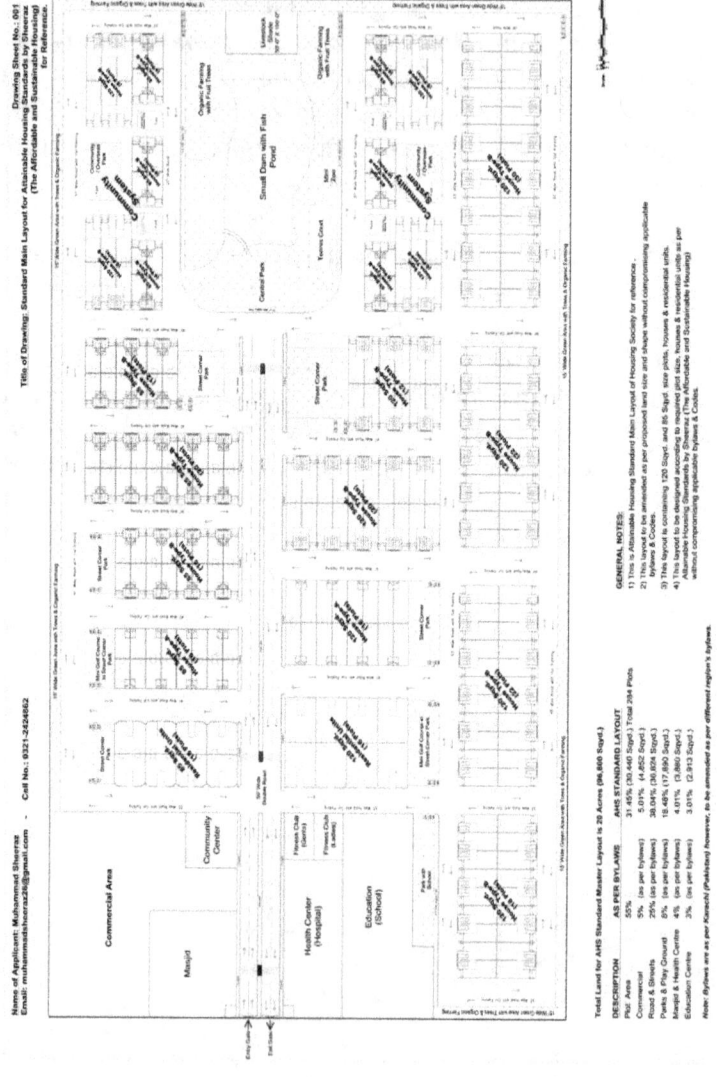

PROPOSED SINGLE STORY 120 SQYD. LAYOUT
(DRAWING SHEET No.: A-120)

Name of Applicant: Muhammad Sheeraz
Email: muhammadsheeraz26@gmail.com
Cell No.: 0321-2424862

Drawing Sheet No.: A-120

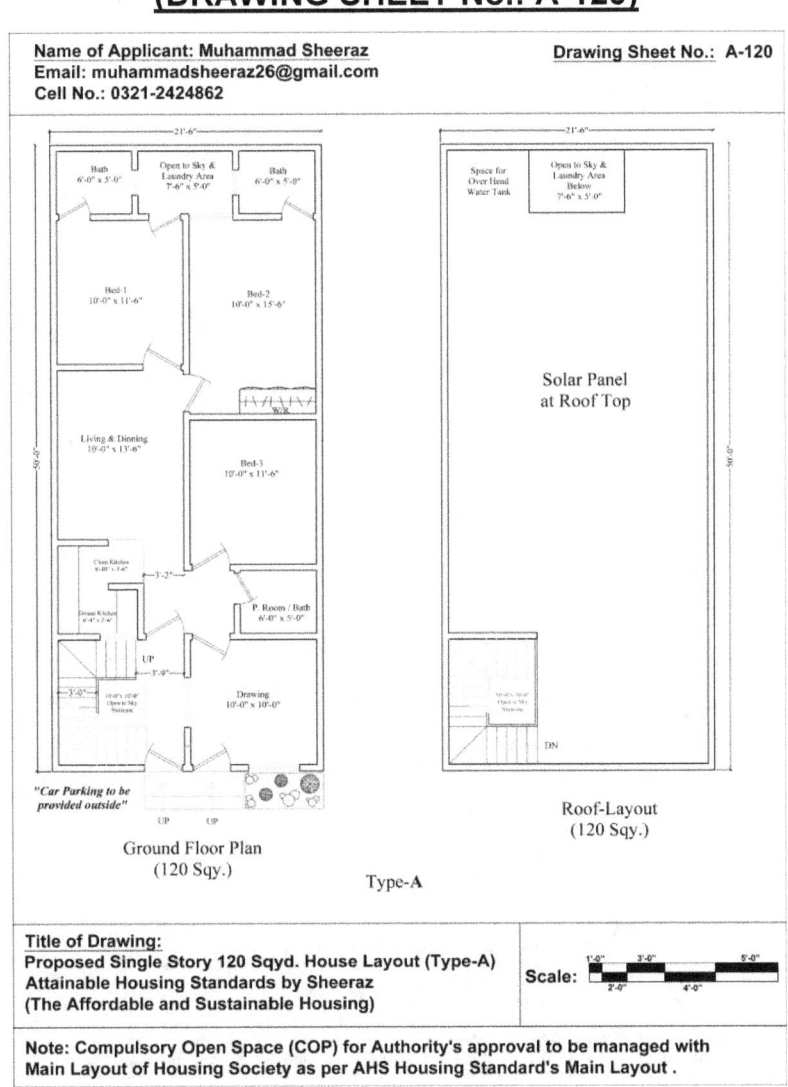

Ground Floor Plan
(120 Sqy.)

Type-A

Roof-Layout
(120 Sqy.)

Title of Drawing:
Proposed Single Story 120 Sqyd. House Layout (Type-A)
Attainable Housing Standards by Sheeraz
(The Affordable and Sustainable Housing)

Scale:

Note: Compulsory Open Space (COP) for Authority's approval to be managed with Main Layout of Housing Society as per AHS Housing Standard's Main Layout .

PROPOSED SINGLE STORY 120 SQYD. LAYOUT

(DRAWING SHEET No.: B-120)

Name of Applicant: Muhammad Sheeraz
Email: muhammadsheeraz26@gmail.com
Cell No.: 0321-2424862

Drawing Sheet No.: B-120

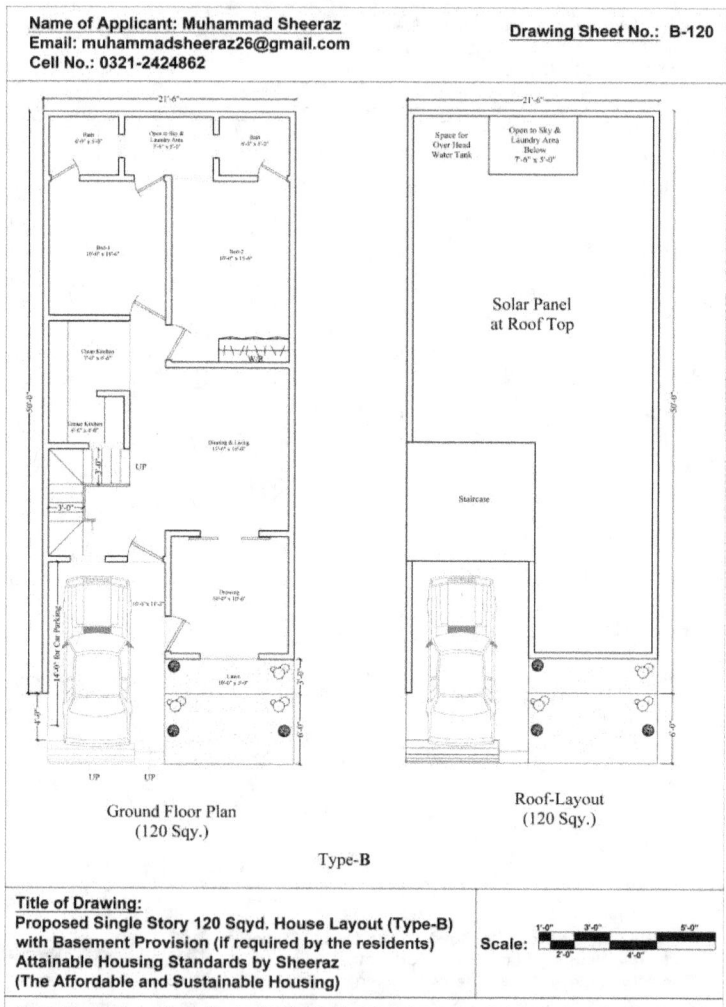

Ground Floor Plan
(120 Sqy.)

Roof-Layout
(120 Sqy.)

Type-**B**

Title of Drawing:
Proposed Single Story 120 Sqyd. House Layout (Type-B)
with Basement Provision (if required by the residents)
Attainable Housing Standards by Sheeraz
(The Affordable and Sustainable Housing)

Scale:

Note: Compulsory Open Space (COP) for Authority's approval already managed however,
further open space to be provided as per AHS Housing Standard's Main Layout .

PROPOSED RESIDENTIAL UNITS 120 SQYD. LAYOUT
(DRAWING SHEET No.: U-120)

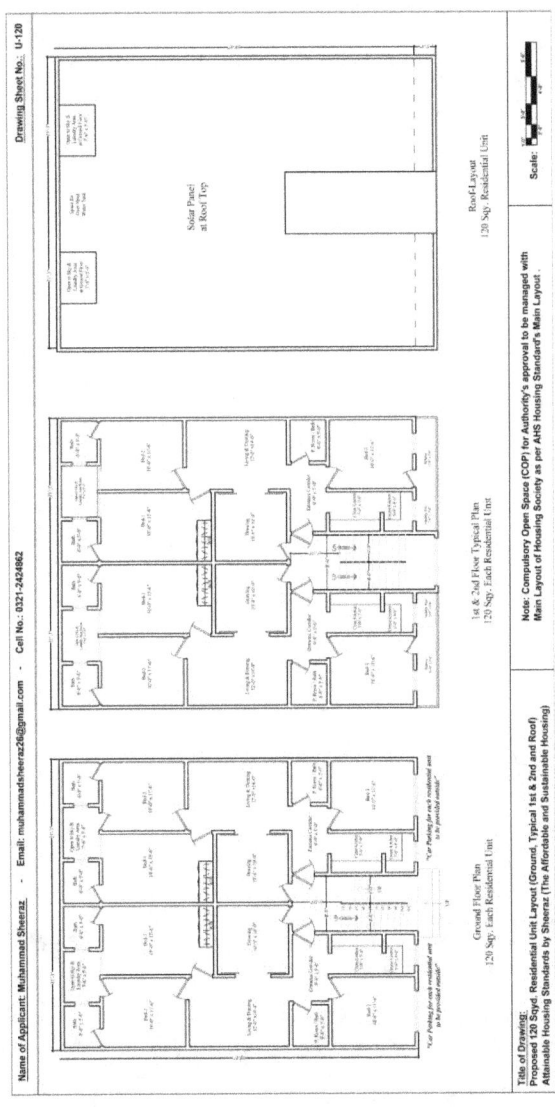

PROPOSED SINGLE STORY 85 SQYD. LAYOUT
(DRAWING SHEET No.: A-85)

Name of Applicant: Muhammad Sheeraz
Email: muhammadsheeraz26@gmail.com
Cell No.: 0321-2424862

Drawing Sheet No.: A-85

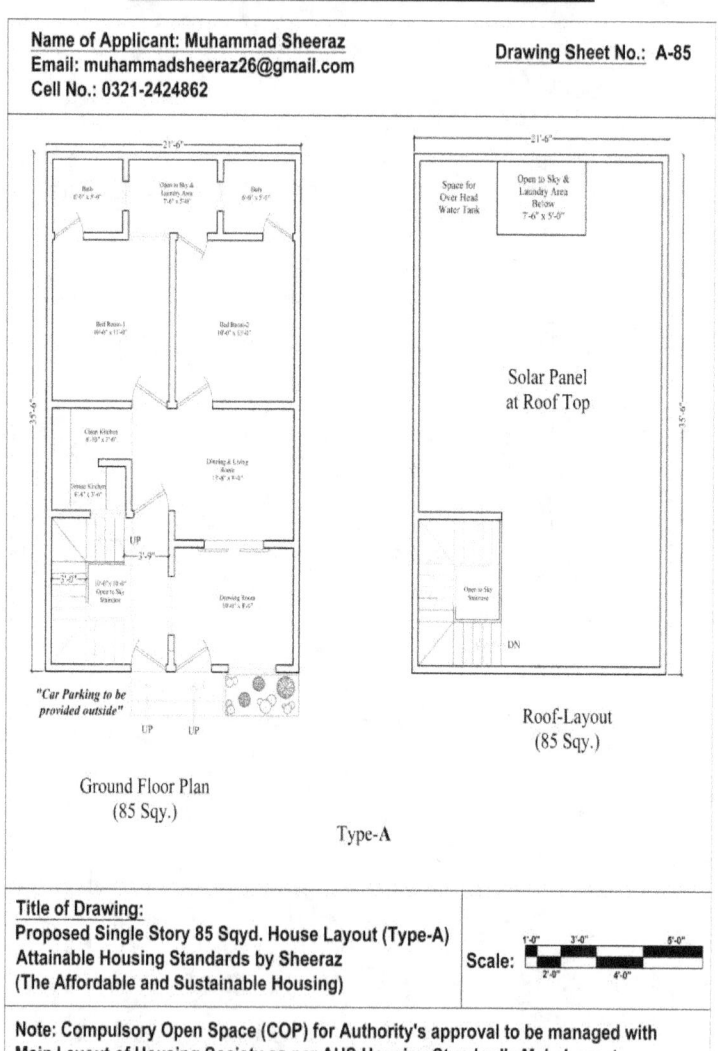

Ground Floor Plan
(85 Sqy.)

Roof-Layout
(85 Sqy.)

Type-A

Title of Drawing:
Proposed Single Story 85 Sqyd. House Layout (Type-A)
Attainable Housing Standards by Sheeraz
(The Affordable and Sustainable Housing)

Scale:

Note: Compulsory Open Space (COP) for Authority's approval to be managed with
Main Layout of Housing Society as per AHS Housing Standard's Main Layout.

PROPOSED SINGLE STORY 85 SQYD. LAYOUT
(DRAWING SHEET No.: B-85)

Name of Applicant: Muhammad Sheeraz
Email: muhammadsheeraz26@gmail.com
Cell No.: 0321-2424862

Drawing Sheet No.: B-85

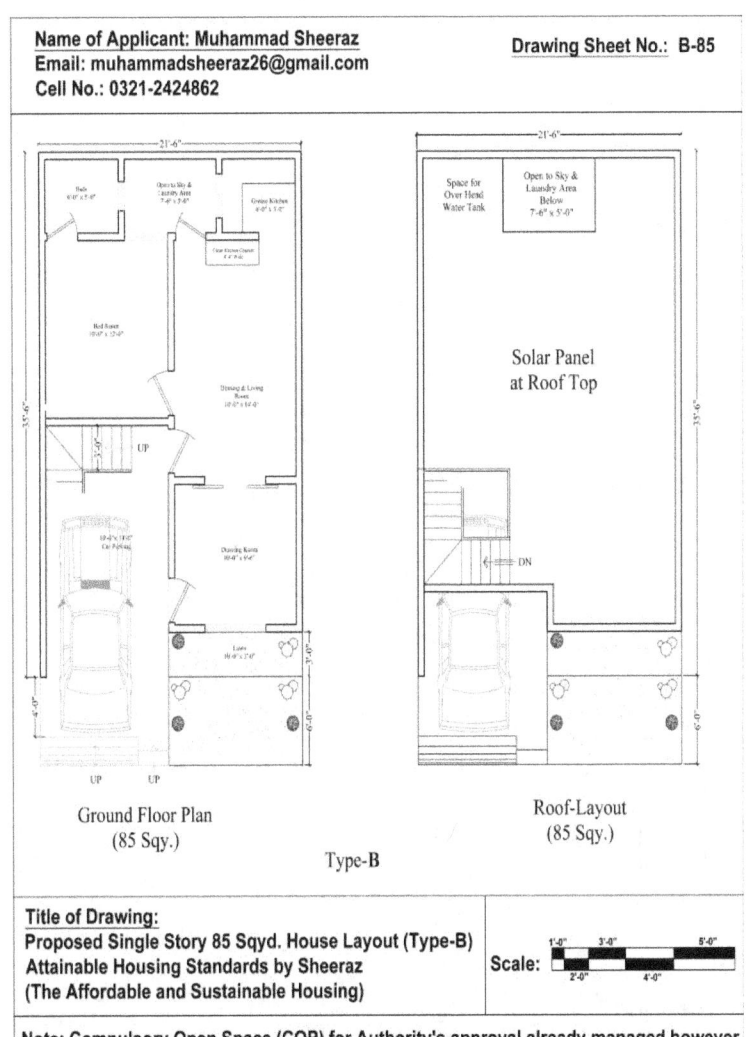

Ground Floor Plan
(85 Sqy.)

Roof-Layout
(85 Sqy.)

Type-B

Title of Drawing:
Proposed Single Story 85 Sqyd. House Layout (Type-B)
Attainable Housing Standards by Sheeraz
(The Affordable and Sustainable Housing)

Scale:

Note: Compulsory Open Space (COP) for Authority's approval already managed however, further open space to be provided as per AHS Housing Standard's Main Layout .

PROPOSED RESIDENTIAL UNITS 85 SQYD. LAYOUT
(DRAWING SHEET No.: U-85)

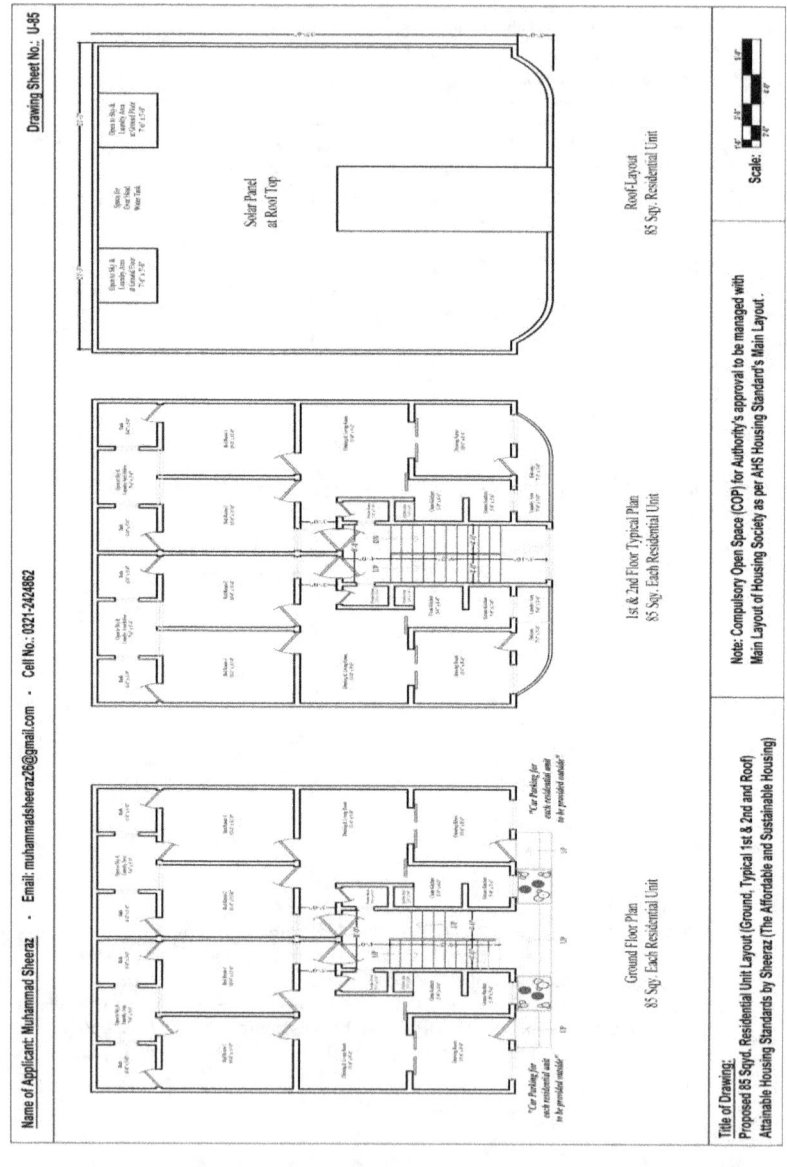

Thanks for reading!
Please provide your reviews
Your feedback is very important and to be
incorporated in next versions with your name
Let me know what you thought!

Reviews are extremely helpful, thank you for taking the time to support my work and me. Do not forget to share your reviews and encourage others to read:
"Attainable Housing Standards by Sheeraz"

DON'T FORGET TO KEEP IN TOUCH FOR
"ATTAINABLE HOUSING STANDARD'S
CERTIFICATION"
AND
"TO DEVELOP THE PROJECTS"
WITH
ATTAINABLE HOUSING (PRIVATE) LIMITED
TO ACHIEVE THE AFFORDABLE AND
SUSTAINABLE HOSUING

Attainable Housing Standards compliance certification, franchises for the Standards Certification services, business opportunities as project development from all nations are always welcome

FOR UPDATES,
MORE INFORMATION,
NEW RELEASES
AND OTHER GREAT READS
muhammadsheeraz26@gmail.com
Call & WhatsApp: +92 321 242 4862
www.ahspakistan.info

www.ingramcontent.com/pod-product-compliance
Lightning Source LLC
Chambersburg PA
CBHW070514220526
45467CB00002B/651